Surrounded By Numbers: Early Math Workbook

Martin's Children's Books LLC
www.martinchildrensbooks.com

Surrounded By Numbers: Early Math Workbook@ Latoshia Martin

All rights reserved. This book or parts thereof may not be reproduced in any form, stored in a retrieval system, or transmitted in any form by any means - electronic, mechanical, photocopy, recording, or otherwise-without prior written permission of the publisher, except as provided by the United States of America copyright law.

Printed in the United States of America

First Edition: November 2021

Surrounded By Numbers
Early Math Workbook

Hi Young Mathematician,

This workbook was created just for you. Use this book to learn to count from 0-100.

This Book Belongs To:

_____

MY NUMBER CHART

1	2	3	4	5	6	7	8	9	10
11	12	13	14	15	16	17	18	19	20
21	22	23	24	25	26	27	28	29	30
31	32	33	34	35	36	37	38	39	40
41	42	43	44	45	46	47	48	49	50
51	52	53	54	55	56	57	58	59	60
61	62	63	64	65	66	67	68	69	70
71	72	73	74	75	76	77	78	79	80
81	82	83	84	85	86	87	88	89	90
91	92	93	94	95	96	97	98	99	100

DIRECTIONS: TRACE THE WORDS AND NUMBERS BELOW.

zero

0 0 0 0

0 0 0 zero

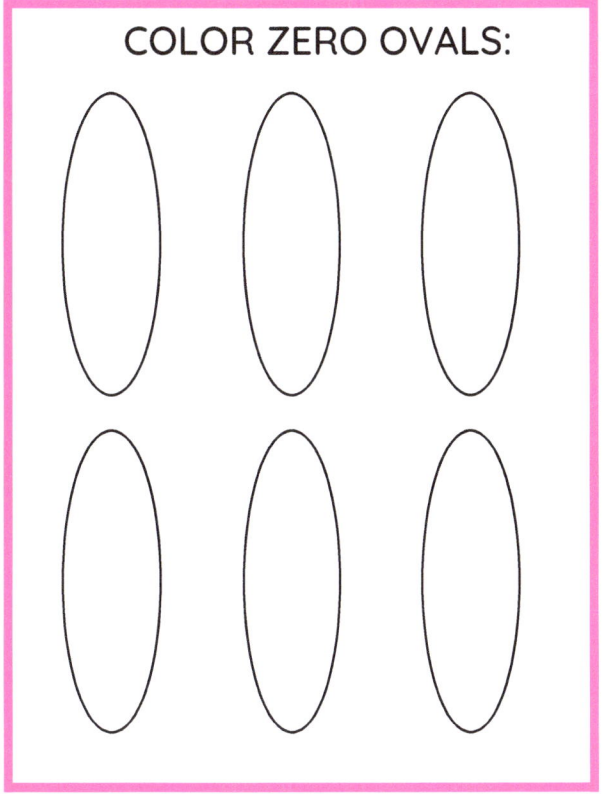

COLOR ZERO OVALS:

CIRCLE THE ZEROES:

5	0	3
1	0	2
0	5	4
4	0	6
0	6	1

DIRECTIONS: TRACE THE WORDS AND NUMBERS BELOW.

one

1 1 1 1 1

1 1 1 1 1 one

COLOR ONE STAR:

CIRCLE THE ONES:

1	3	5
4	2	1
6	1	8
1	7	1
2	1	4

DIRECTIONS: TRACE THE WORDS AND NUMBERS BELOW.

2

two

2 2 2 2

2 2 2 2 two

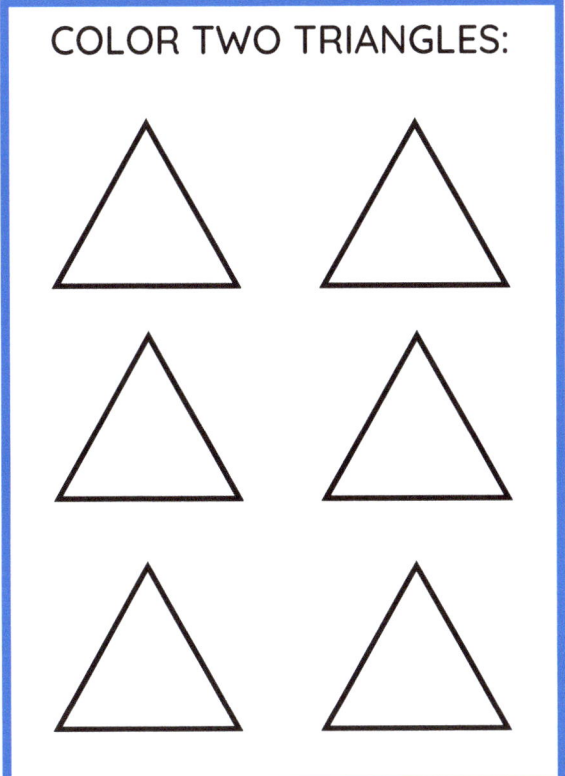

COLOR TWO TRIANGLES:

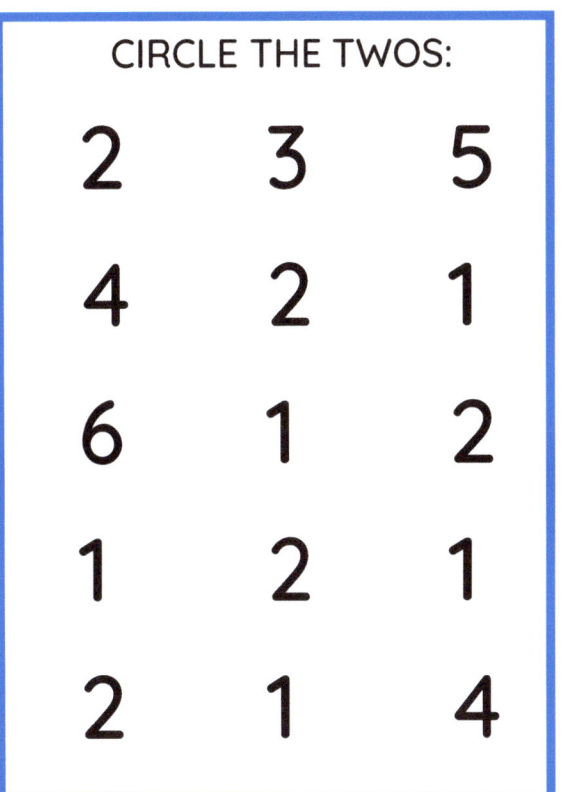

CIRCLE THE TWOS:

2	3	5
4	2	1
6	1	2
1	2	1
2	1	4

DIRECTIONS: TRACE THE WORDS AND NUMBERS BELOW.

3

three

3 3 3 3

3 3 3 three

COLOR THREE SQUARES:

CIRCLE THE THREES:

2	3	5
4	2	3
3	1	2
3	5	1
6	3	4

DIRECTIONS: TRACE THE WORDS AND NUMBERS BELOW.

four

4 4 4 4

4 4 4 4   four

COLOR FOUR HEARTS:

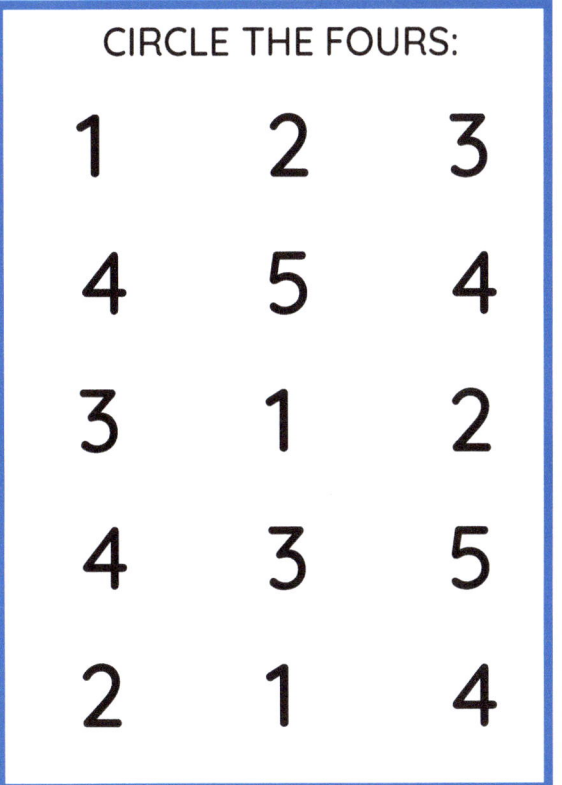

CIRCLE THE FOURS:

1 2 3
4 5 4
3 1 2
4 3 5
2 1 4

DIRECTIONS: TRACE THE WORDS AND NUMBERS BELOW.

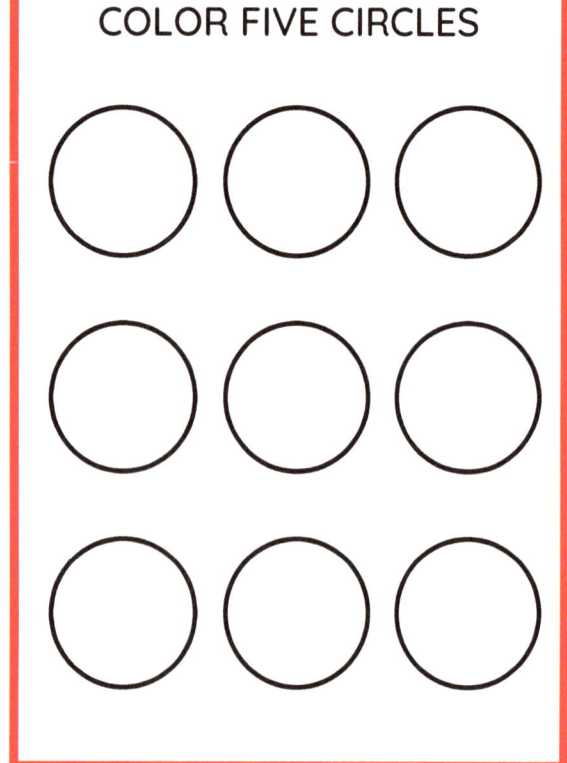

COLOR FIVE CIRCLES

CIRCLE THE FIVES

5	2	3
1	5	4
3	5	2
4	3	5
2	5	4

Number Matching

Draw a line matching the numerals and the numbers represented with the hands.

Number Matching

Draw a line matching the numerals and the numbers represented with the hands.

DIRECTIONS: TRACE THE WORDS AND NUMBERS BELOW.

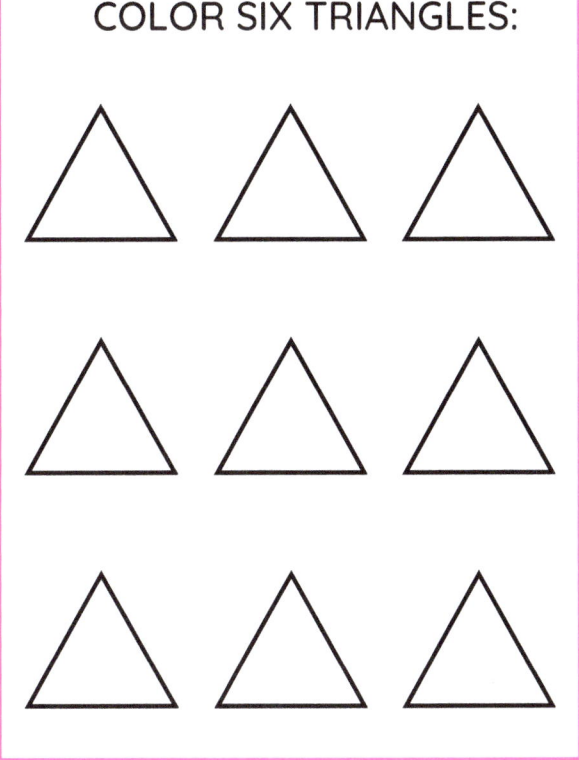

COLOR SIX TRIANGLES:

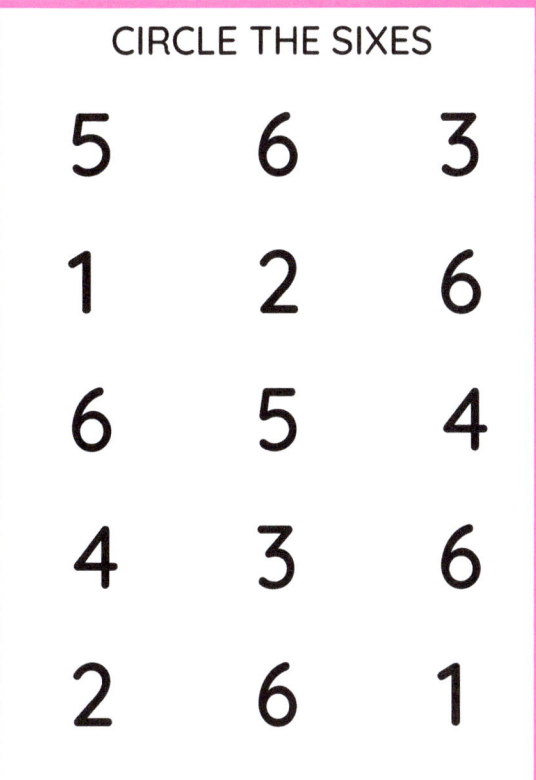

CIRCLE THE SIXES

5	6	3
1	2	6
6	5	4
4	3	6
2	6	1

DIRECTIONS: TRACE THE WORDS AND NUMBERS BELOW.

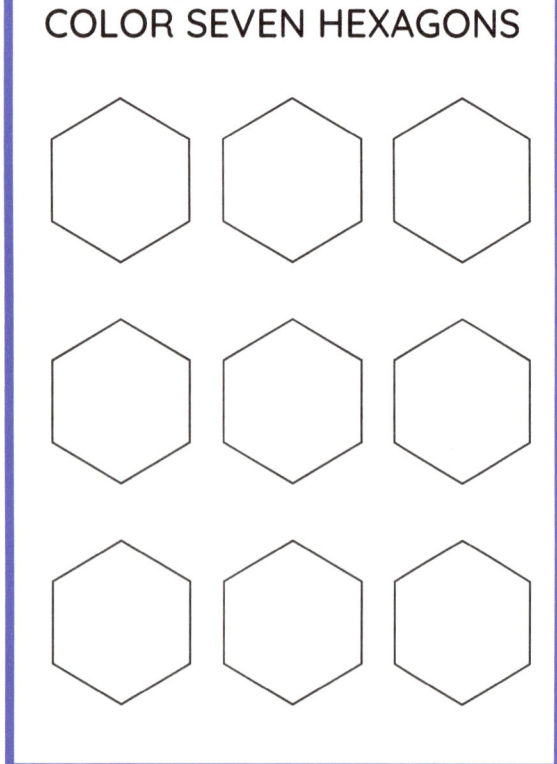

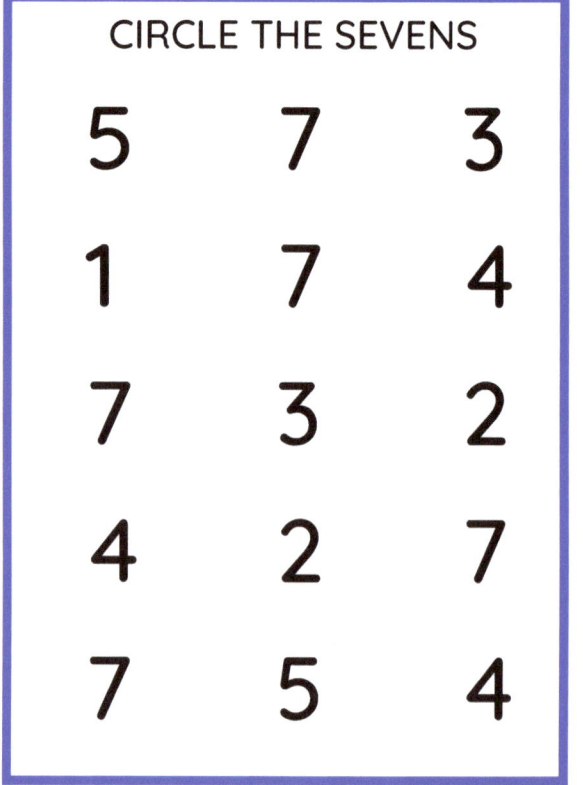

DIRECTIONS: TRACE THE WORDS AND NUMBERS BELOW.

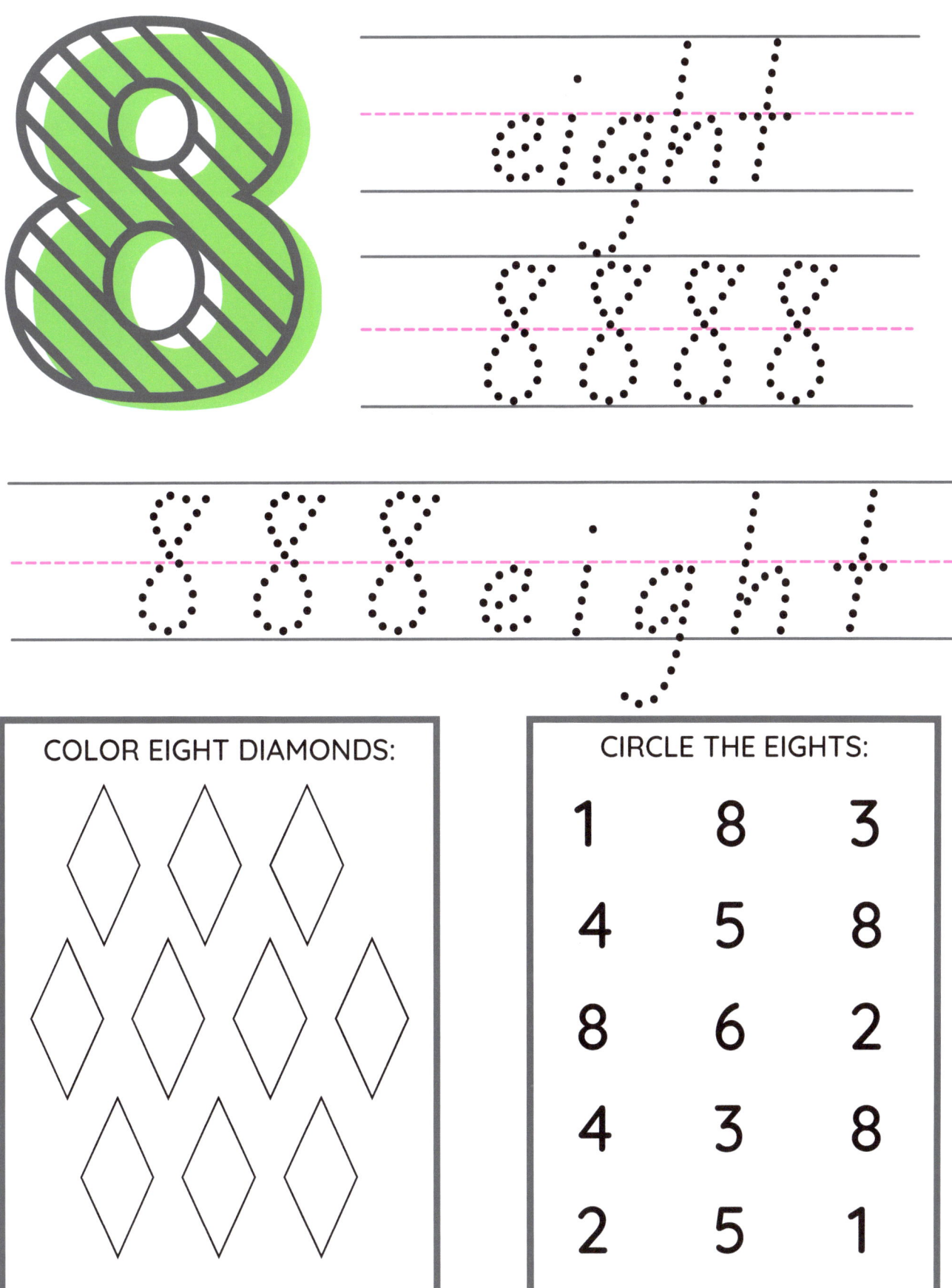

DIRECTIONS: TRACE THE WORDS AND NUMBERS BELOW.

nine

9 9 9 9

9 9 9 9 nine

COLOR NINE HATS:

CIRCLE THE NINES:

9	8	6
4	7	9
8	6	9
4	9	8
9	5	1

DIRECTIONS: TRACE THE WORDS AND NUMBERS BELOW.

10   ten
      10   10
10 10 ten

COLOR TEN SOCKS:

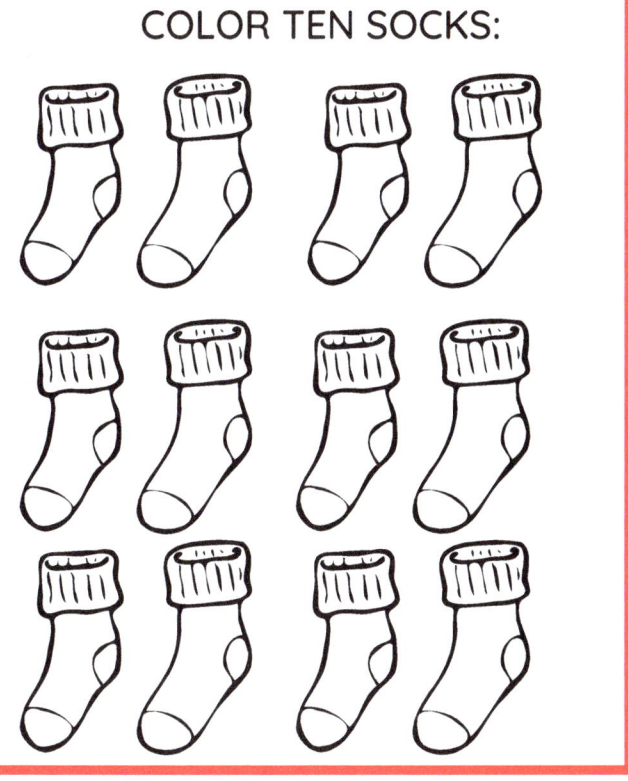

CIRCLE THE TENS:

10	3	9
4	2	1
6	1	8
10	7	10
2	1	4

I CAN WRITE NUMBERS

Practice writing your numbers. Trace the first three then complete the row on your own.

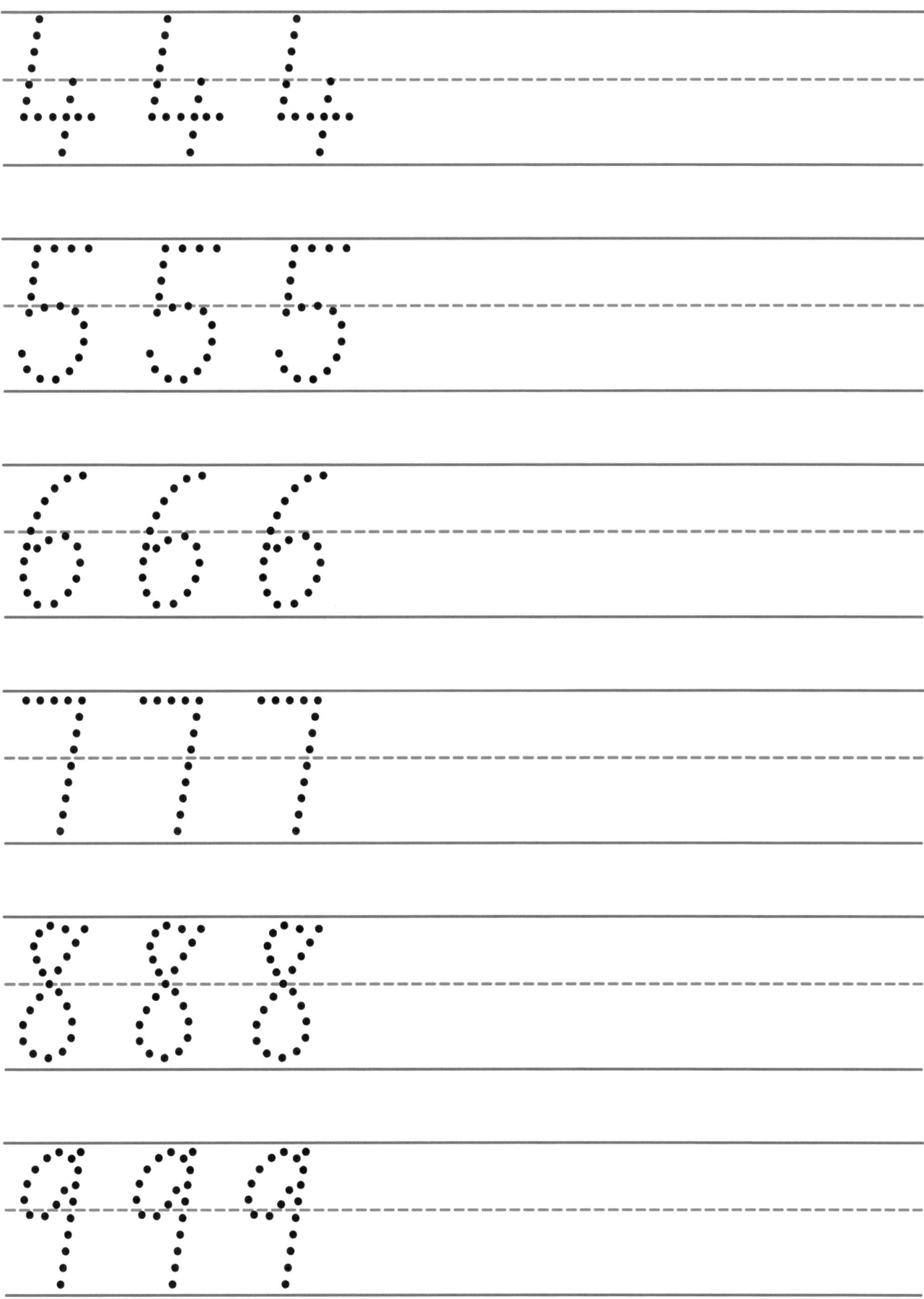

Number Values

Color the number that has the bigger value

5 8	4 5
6 0	3 2
9 5	4 7
8 2	6 1

Count & Color

Count and color the exact number

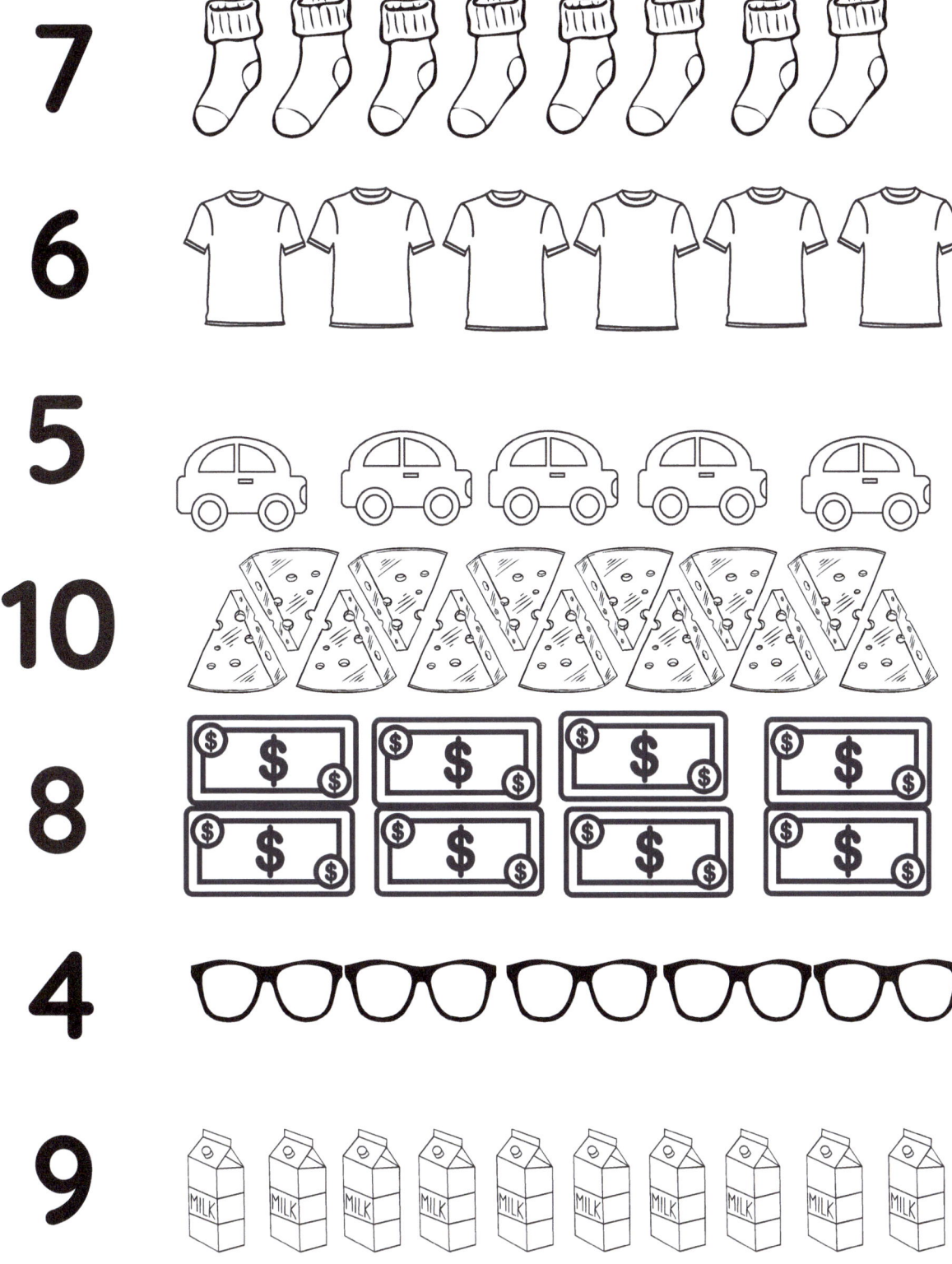

Counting 1 to 20

READ IT

1	2	3	4	5	6	7	8	9	10
11	12	13	14	15	16	17	18	19	20

TRACE IT

1	2	3	4	5	6	7	8	9	10
11	12	13	14	15	16	17	18	19	20

WRITE IT

I Can Count!

Write the missing numbers using the number line for help, if needed.

| 8 | 9 | 10 |

| 11 | ___ | ___ |

| 2 | ___ | ___ |

| 15 | ___ | ___ |

| 13 | ___ | ___ |

| 7 | ___ | ___ |

| 4 | ___ | ___ |

| 1 | ___ | ___ |

| 18 | ___ | ___ |

| 10 | ___ | ___ |

| 6 | ___ | ___ |

| 14 | ___ | ___ |

1 2 3 4 5 6 7 8 9 10 11 12 13 14 15 16 17 18 19 20

Before and After

Write the numbers that come before
and after the number in the middle.

8	9	10			7	
	6				11	
	3				3	
	5				5	
	12				4	
	18				2	

Ordinal Numbers

First - 1st	Eleventh - 11th
Second - 2nd	Twelfth - 12th
Third - 3rd	Thirteenth - 13th
Fourth - 4th	Fourteenth - 14th
Fifth - 5th	Fifteenth - 15th
Sixth - 6th	Sixteenth - 16th
Seventh - 7th	Seventeenth - 17th
Eighth - 8th	Eighteenth - 18th
Ninth - 9th	Nineteenth - 19th
Tenth - 10th	Twentieth - 20th

Ordinal Number Word Search

Find the word form of the number.

```
F I F T E E N T H J T
F T W E L F T H B H H
E L E V E N T H M I I
P F H N A N A A S V R
P O A O T K O S E T T
F U L E U H N A V I E
I R G G T H I E W E
F T G X S E C O N D N
T H I R D Y C A T T T
H S I N R E I G H T H
X F O U R T E E N T H
```

1st	6th	11th
2nd	7th	12th
3rd	8th	13th
4th	9th	14th
5th	10th	15th

Surrounded By Numbers
by: Latoshia Martin

Ordinal Numbers

Draw a line to match the numbers on the left with its word form on the right..

4th	First
10th	Third
1st	Tenth
2nd	Sixth
8th	Second
3rd	Fourth
7th	Seventh
6th	Eighth
9th	Ninth
5th	Fifth

Coloring Ordinal Numbers

Directions:
- Color the 2nd building red
- Color the 5th building yellow
- Color the 9th building blue
- Color the 1st building green.
- Color the 6th building orange

What's missing?

Practice writing your numbers 1 - 100 by tracing the numbers in order below when you come to an empty box, write in the missing number.

1	2		4	5	6		8	9	10
11	12	13	14		16	17	18		20
21		23	24	25	26		28	29	30
	32	33	34	35	36	37		39	40
41	42	43			46	47	48	48	50
51	52	53	54	55		57	58	59	
61	62		64	65	66	67	68		70
	72	73	74	75	76		78	79	80
81		83	84	85	86	87		89	90
91	92		94	95	96	97	98		100

I Can Write!

Practice writing your numbers on your own..

I Can Write!

Practice writing your numbers on your own..

I Can Write!

Practice writing your numbers on your own..

LETTER & NUMBER SORTING

Practice writing the letters and numbers below by tracing over them with your pencil. When finished, cut out the numbers and letters below and sort affix them on the proper jars.

A	4	c	9	k	5	r	3	D
6	H	8	Q	2	w	f	M	7

Tear/Cut Out The Last Page

Tape it to your wall.

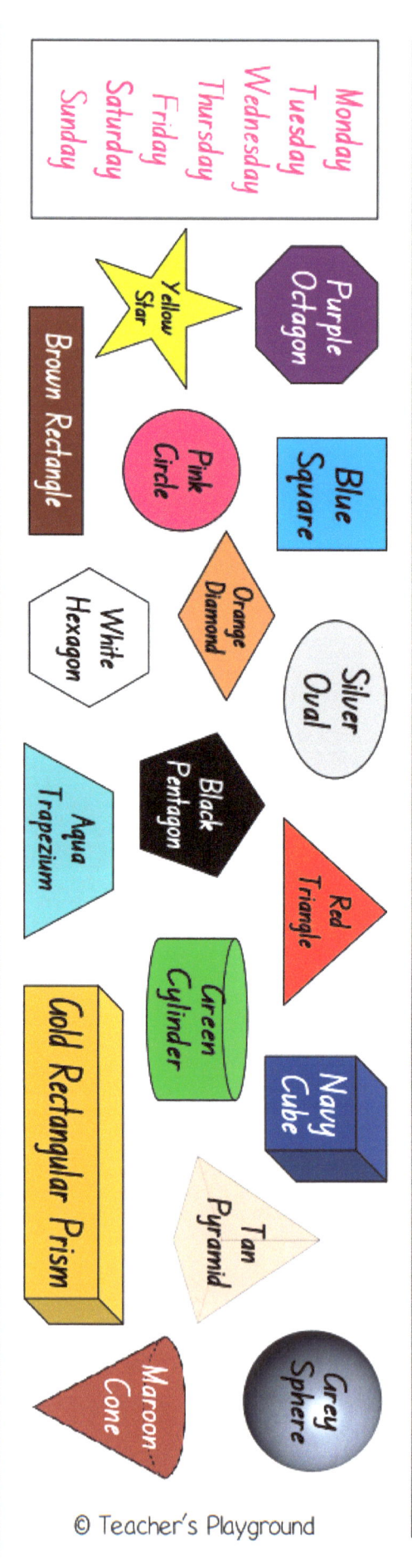

 www.ingramcontent.com/pod-product-compliance
Lightning Source LLC
Chambersburg PA
CBHW042249100526
44587CB00002B/73